DIAMONDS

Vicky Paterson

FIREFLY BOOKS

A Firefly Book

Published by Firefly Books Ltd. 2005

First printing

Publisher Cataloging-in-Publication Data (U.S.)

Paterson, Vicky.
    Diamonds / Vicky Paterson
[112] p. : col. photos. ;   cm.
Includes index.
Summary: A guide to the science and glamour of diamonds. Covers topics such as, why diamonds are highly prized and their properties; how rocks are transformed into jewels; the role that diamonds play in modern technology and medicine; and famous diamonds and their legends.
ISBN 1-55407-116-X
1. Diamonds, Artificial. 2. Diamond thin films.
I. Title. 666.88 dc22   TP873.5.P3847 2005

Library and Archives Canada Cataloguing in Publication

Paterson, Vicky, 1974 –
    Diamonds / Vicky Paterson
Includes index.
ISBN: 1-55407-116-X
    1. Diamonds. I. Title.
TS753.P38 2005   553.8'2   C2005-903028-3

Published in the United States by
Firefly Books (U.S.) Inc.
P.O. Box 1338, Ellicott Station
Buffalo, New York 14205

Published in Canada by
Firefly Books Ltd.
66 Leek Crescent
Richmond Hill, Ontario L4B 1H1

Printed in England

To my dear Uncle Salah

Front cover: The De Beers Millennium Star © De Beers LV

Spine: Marquise-cut diamond © The Diamond Trading Company Limited

Back cover from top to bottom:
Round-cut diamond © Aurora Gems/Photo Tino Hammid
Radiant-cut diamond © Aurora Gems/Photo Robert Weldon
Marquise-cut diamond © The Diamond Trading Company Limited
Oval-cut diamond © Aurora Gems/Photo Robert Weldon

Designed by Mercer Design

Contents

CHAPTER ONE

Desire

DIAMONDS ARE THE MOST SOUGHT-AFTER AND GLAMOROUS GEMS IN THE WORLD.

THEIR STORY TAKES US FROM COASTS TO DESERTS AND EVEN INTO SPACE.

Every year, many millions of carats of diamonds are dug out of the ground or extracted from the beds of rivers and seas. These precious stones then begin an epic journey to markets around the world. Some are bought by the rich and famous for their beauty and status. Others are destined for science and industry, where their unique physical properties are harnessed. The market for uncut diamonds is worth an estimated £5.5 billion (US$11 billion) annually. That's a lot for a stone made from nothing but carbon, just like the graphite in a pencil lead. But when you explore the story of diamonds, and dig beneath the surface from their sparkle to their science, it becomes clear why.

Diamonds are notoriously difficult to find, a fact that makes them even more desirable. Most were formed deep inside the Earth billions of years ago. Diamonds are even more ancient than the rocks in which they occur. But these rocks lie hidden in long-dead volcanoes that over the years have been worn away to lie flat and concealed on or under the landscape. Diamonds are sometimes eroded out of the volcanic rock and washed downstream to surrounding river beds or even to the coast. It takes expertise and patience to find them, not to mention money.

LEFT Tonnes of rock must be crushed to find even the smallest diamonds. This fact makes the De Beers Millennium Star, with a weight of 203.04 carats, seem even more magnificent.

LEFT This watercolour
c. 1616–1617 shows the
future Indian Emperor,
Shah Jahan, admiring a
diamond and emerald turban
ornament.

Our enduring desire for diamonds began at least 4000 years ago in India, where some of the highest-quality stones ever found originate. The country was a major source of diamonds until 1725, when they were discovered in Brazil. But it was the huge finds in South Africa from 1866 onwards that caught the world's attention, sparking a diamond rush that spread like wildfire.

Diamonds are now mined on every continent except Europe and Antarctica. They were most recently discovered in Canada, in 1991. And the major discovery in 1979 of the Argyle Mine in Western Australia has resulted in Australia becoming one of the world's largest producers by volume. The money made from the sale of the precious stones is, by contrast, far more concentrated. Around half of all rough or unprocessed diamonds dug out of the ground end up being sold through the South African firm De Beers.

The most desirable diamonds are made into stunning jewellery, but in the ground they may seem slightly less amazing. It is only when cut and polished that they are transformed into beautiful jewels, dazzling onlookers with a characteristic 'fire'. Few people realise that diamonds are not always clear or 'white', but come in a variety of colours – pinks, yellows, greens, blues and browns. To many, the incomparable beauty of this sparkling radiance alone justifies the hefty price tag.

LEFT An elaborate turban ornament featuring diamonds in a gold setting with emerald droplets, now held at the Musée Guimet, Paris.

RIGHT This superb example of an Indian-influenced design was crafted by Cartier in 1931. The unusually large diamonds make it particularly appealing.

Diamonds are the heavyweight jewels of the rich and famous. Film stars and musicians can make the headlines as much for the jewellery they wear as the work they do. Diamonds make the ultimate statement about wealth and status, where 'bling' often speaks louder then words. And they have also secured a place in the hearts of many thousands of couples who buy engagement rings. However romantic the setting or well-meaning the delivery, can a ruby or other gem cut it in quite the same way?

But the reach of diamonds goes far beyond the weighted wrists of hip-hop artists or the fingers of newly-weds. Industry values diamonds, too. Harder than steel, diamond is used to coat heavy-duty cutting tools and saws. Its ability to spread heat and resistance to scratching, enabling super-transparency, can be applied in many technologies. It can withstand high-frequency, high-voltage conditions such as in communications satellites, and is also resistant to radiation. One day, diamonds might also be used in machines so small they can be injected into our bloodstream to monitor our health. Scientists are even looking at ways to use diamond to grow new nerve cells and astronomers are finding them in the stars. Such is the demand for them, synthetic diamond material is now being grown in the laboratory. Thousands are made for industry each year and gem-quality synthetics for use in jewellery are already available.

RIGHT From Tiffany, a Lucida® cut diamond-and-platinum engagement ring, balanced on a wedding band.

So how much do gem-quality diamonds cost for the average buyer? Each stone is priced almost entirely on looks. How clear it is, its colour, how few spots of impurities can be seen, how it's cut and how much it sparkles all affect the final price tag. The clearer it is, and the more dazzling, the more a buyer can expect to pay for it. The diamond's weight is also taken into consideration, so only buyers with serious money to spend can be too choosy about the dimensions of their 'rock'. High-street stores sell thousands of affordable diamonds every year, where for instance a small gold pendant embedded with a tiny piece of diamond costs about £30 (US$60). Those wanting a more significant stone, may find their pulse rate quickening when they realise that even a relatively small diamond, less than one carat, of average quality, may cost them hundreds if not thousands of pounds. Even a piece the size of a cherry stone can set a buyer back tens of thousands of pounds. Some sell for millions.

LEFT Not one for the budget conscious, this exquisite Graff white diamond necklace displays multi-shaped stones and culminates in a fancy light pink pear-shaped diamond drop.

RIGHT Flawless pear-shaped earrings from top jewellery house Graff.

RIGHT AND BELOW Diamonds are virtually impossible to trace to source once cut.

But there are risks to the reputation of the diamond trade. Although nearly all rough diamonds are from legitimate sources, some have been sold to finance civil wars. These 'blood' or 'conflict' diamonds, as they are known, have fuelled unrest and human rights abuses that have decimated African countries such as Liberia, Sierra Leone, the Democratic Republic of Congo and Angola. Diamonds in these countries are easily extracted from residual deposits in river beds, without the need for heavy machinery, and are easily transported as they are small and light. Once mixed with diamonds from other sources they are almost impossible to trace.

The key to protecting the reputation of this unique product lies with every mining company, government, trader, jewellery retailer and consumer, to insist diamonds can be proven to be from a legitimate source at every stage of the chain. In 2000, major diamond trading and producing countries, representatives of the diamond industry and civil society, met in South Africa to work out how to tackle the problem of conflict diamonds. This marked the beginning of a three-year negotiating process to set up an international diamond certification scheme. Finally, in 2003 this culminated in the establishment of the Kimberley Process Certification Scheme, endorsed by the United Nations General Assembly and the United Nations Security Council. Forty-three participants, including the European Union, have signed up to it so far, representing 99.8% of the market. The scheme requires that all participating countries agree to only trade in rough

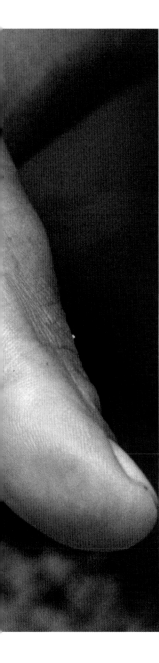

diamonds with other participating countries, and to establish national import and export controls to keep conflict diamonds out of the legitimate trade. It guarantees the origins of diamonds, and gives written proof they have not been used for illicit purposes.

The Kimberley Process is a major step in the right direction, but still has a way to go to ensure diamonds can never again be used to fund destructive activities. Gaps remain in the system, which continue to harbour conflict-diamond traders. The Republic of Congo (not to be confused with its neighbour the Democratic Republic of Congo) was suspended from trading its diamonds in 2004, as it failed to meet the Kimberley Process minimum requirements. How effective this has been is hard to measure, but it sends a clear message that times are changing.

The irony is that many of the countries producing some of the world's most expensive gems are themselves economically poor. This may contribute to a politically volatile environment in which diamonds are once again exploited by rebel groups and terrorists.

It is clear that the diamond story has many facets. Far from being merely 'a girl's best friend', these beautiful gems have a long history, and a long future. How exactly diamonds will change our technology, and how their trade will be successfully controlled no one quite knows. But our desire for them will last a lifetime.

LEFT Some of the countries most rich in diamonds still battle with poverty, partly because profits are drained by civil unrest.

CHAPTER TWO

The Hunt

DIAMOND DEPOSITS ARE SCATTERED ACROSS MANY COUNTRIES AROUND THE WORLD.

FROM NAMIBIA'S SANDY COAST TO THE ANCIENT BEDROCK OF SIBERIA.

The Earth does not give diamonds up without a fight. The hunt for these precious and sought-after stones is fraught with difficulty and chance. But with a market worth about £5.5 billion (US$11 billion) a year, and rising, the lure for some is irresistible. The modern-day hunt for diamonds began on the banks of South Africa's Orange River. It was here in 1866 that the country's first diamond discovery was made by Erasmus Jacobs, a 15-year-old farmer's son – humble beginnings for a business that was to transform diamond history. In 1869 the first primitive mines were established in South Africa, and so began the great diamond rush of the nineteenth century. Eager prospectors were allocated a plot of land each on which to look for diamonds. It was by amalgamating the

LEFT After South Africa's diamond rush of 1869, officials controlled the flood of prospectors by allocating each a 2.9 m² (31 ft²) plot on which to dig.

claims on a farm once owned by the De Beers brothers, that Cecil John Rhodes, an English entrepreneur, built the largest diamond mining company in the world: De Beers. Diamonds have since been found across the globe. About 25% of the annual market comes from Botswana, with most of the rest from South Africa, Australia, Russia, the Democratic Republic of Congo and Canada.

Only by exploring the astonishing way in which diamonds are created and brought to the surface can the thrill of finding them be realised fully. Diamonds form deep within the Earth. Most were created in rocks at least 2.5 billion years old, at about 140-200 km (87-124 miles) underground. Some form even deeper. The intense pressures and great heat at these depths cause the scarce carbon atoms present to become tightly packed together, forming the distinctive crystal structure that so defines a diamond. There they would remain buried, were it not for occasional and particularly violent volcanic eruptions. These eruptions originate beneath the zone of diamond formation, far deeper than most eruptions. They punch through the old rocks containing the diamonds, which are then rapidly carried to the surface in the magma, or molten rock, to explode at the surface like a popped bottle of champagne.

The magma systems which feed volcanoes such as Vesuvius or Mount St Helens are only about 60 km (37 miles) deep. Diamonds, however, form at three times that depth. These deep volcanoes produce a rare type of rock called kimberlite, named after the Kimberley area in South Africa where it was first defined. It is in kimberlite, and a similar rare volcanic rock called lamproite, that diamonds are found.

The journey to the Earth's surface is fraught with risk. Diamonds may be absorbed in the magma. If they cool too slowly on the way up, they may turn to graphite. And any contact with oxygen may turn the diamonds into puffs of carbon dioxide gas. When the erupted magma cools, it forms rock with the surviving diamonds scattered through it. This rock is contained in a narrow, often carrot-shaped 'pipe' or in a thin fissure. Over hundreds of millions of years the surface of the pipe slowly erodes away, leaving little visible trace, and often it becomes buried beneath younger rocks. In the process, some of the diamonds are washed downstream and accumulate in river beds or on shorelines.

Alluvial diamond deposits found in river beds are generally small, although there are exceptions. The Democratic Republic of Congo is a big producer of diamonds, most of which are alluvial. They are easier to extract, and in theory could be found by anyone with the patience to sift for hours through sand and gravel. Alluvial production worldwide amounts to approximately 6 tonnes (5.9 tons), about 30 million carats (diamonds are measured in carats; 1 carat = 0.2 g/0.007 oz). The estimated total diamond production for 2004 was 142 million carats, and roughly one-fifth of that was alluvial.

Diamonds washed down to the coast can also be moved by the continuous action of currents further out to sea, where they accumulate on the seafloor. These marine deposits do not escape the prospectors, as the diamond-bearing sediment is either sucked up through large pipes to ships on the surface or collected by underwater 'tractors'.

RIGHT Not all diamonds are found by crushing tonnes of rock. Some are extracted by sifting gravel downstream from the source, as pictured here on the São Luiz River, the Mato Grosso region of the Amazon.

But onshore mining is where the big money is. Highly mechanised equipment on a huge scale is used to penetrate the kimberlite pipe, hauling out truckloads of rock to be crushed and gleaned for their precious stones. The crushed rock is usually passed through a separator lined with grease, to which diamonds readily stick. Traditionally, the other constituents of the rock are then washed off, leaving the diamond crystals behind. But modern machines can now pick out diamonds from the rubble, using a beam of X-rays that makes individual stones glow. Once a glowing stone is detected, it triggers a jet of air, which deflects the diamonds into a collecting bin.

LEFT De Beers's Finsch mine in South Africa is a relatively big kimberlite pipe, 0.18 km². It has been producing diamonds at a high rate since 1967, and is the most modern underground mine in the world.

Thousands of workers operate the bigger mines, some transporting the kimberlite to be processed, others handpicking the biggest and best rough diamonds from the broken rock. In some cases, workers live at the secure site for six months at a time, requiring only one security search: when they leave. Prospectors dream of finding kimberlite pipes, each a potential treasure trove. But how do you find a kimberlite pipe when it can no longer be seen from the surface, if over the years it has been eroded down and covered up by desert sands or glacial wastes? The methods are numerous and varied, ranging from sensitive satellite imaging from outer space to meticulous soil sampling on the ground.

Kimberlite does not only contain diamonds: it also contains the more common minerals, including certain compositions of both garnets and ilmenite. Soil churned up by burrowing animals such as rabbits and termites is examined for these 'indicator' minerals. If they are found, they may signal that kimberlite is near, which means diamonds might be too.

Even kimberlite hidden under ice doesn't escape the gaze of hungry prospectors. In one recent instance, Canadian prospectors were looking for diamonds in the Northwest Territories, up towards the Arctic Circle. Conditions were harsh. Bears and wolves stalked the icy, wind-whipped tundra, which was largely devoid of people. Suddenly, the prospectors' spirits lifted when they found indicator minerals around the banks of a frozen lake covered in ice 2 m (6½ ft) deep. It was a gamble, but they sent for hi-tech equipment that could detect anomalies in the Earth's geomagnetic field. The results they picked up confirmed the presence not only of rock typical of the area, but also an enormous pipe. They had found kimberlite, and the discovery would unleash Canada on to the diamond scene as a potential world leader.

But it isn't only scientific investigations that reap rewards. 'Hunches' from environmental observations can also lead to the treasure. Prospectors often scout an area from the sky in light aircraft, picking up electromagnetic signals using on-board equipment, or looking for anomalies in the landscape such as irregular dips or unexplained breaks in vegetation. These can sometimes indicate a change in rock under the surface, and sometimes that rock is kimberlite. One story goes that a prospector noticed a group of cows grazing in a large field, but only in one area of it. After closer inspection, the grass in that area was found to be growing on richer soil,

LEFT Prospectors make aerial surveys using electromagnetic equipment to detect kimberlite pipes.

which turned out to lie above a huge kimberlite deposit. Only a relatively few pipes contain diamonds, and some do so to a greater degree than others. But the potential rewards from a diamond-rich pipe are phenomenal.

Once a kimberlite pipe is found it must first be tested for diamonds. It is drilled and cored in order to prove that diamonds are both present, and present in sufficient quantities. This is done on a grid-like basis so that both the size of the deposit and its economic viability can be estimated. The need to establish these factors from the outset becomes apparent when you consider that even some of the richest diamond deposits produce only one carat of diamond for every 3 tonnes (2.9 tons) of rock mined and crushed.

The hunt for diamonds has revealed some quite spectacular deposits. One of the most productive mines, which was later dubbed The Big Hole, produced continually from 1871 until 1914. It was one of the first South

African mines, and was bought by De Beers along with many others. Russia became the first big producer outside of Africa when in 1955 a huge pipe was found in the Sakha region, in north-central Siberia. It was called the Mir (Peace) mine, but is now closed due to flooding. Western Australia's Argyle Mine in the Ellendale area of the Kimberley Plateau, has produced more than 30 million carats of diamonds every year since it began production in 1981.

Finding diamonds isn't always the end of the hunt. Not all diamonds are big money-makers. Discovering gem-quality diamonds, the ones good enough to set in jewellery, is what sets the big players apart from the rest. In 1908, a railroad worker working in a small town on the coast of Namibia, in southwest Africa, plucked a diamond out of the sand. Soon the area was being harvested by workers crawling on their hands and knees. The area has since been heavily mined, on the shore and increasingly out to sea. These diamond beds are not kimberlite, but huge accumulations of sands eroded from across southern Africa, which includes some material eroded from the kimberlite. Namibia produces an exceptional proportion of gem-quality stones, representing 90–95% of its total diamond production.

LEFT Marine deposits, such as those in Namibia, are often rich in high-quality stones, transported in a turbulent journey from source to sea.

So how many diamond deposits are there in the world and will we ever find them all? The simple answer is we don't know. As long as there is a market for diamonds the dramatic hunt for them will continue until this resource is exhausted. Some scientists have suggested that the Earth is no longer making diamonds like it used to. By examining rock from diamond-rich areas, gemmologists have been able to make estimates about the type and age of diamonds we have been mining for the past 150 years. What they found was a surprise. Rather than an even spread of ages, their research indicates there have been three or more distinct generations of diamonds. The first was formed more than three billion years ago, the second just under three billion years ago and the third around one billion years ago. Some researchers suggest that conditions were different then, that the planet was hotter internally, or rock composition was different. But is seems probable that only a very few would form in the Earth's mantle today.

ABOVE LEFT TO RIGHT
'Fancies' – attractively coloured diamonds – are more rare than their 'white' counterparts, but their popularity depends on what is in vogue.

The market for diamonds is as varied as the stones themselves. About 80% of diamonds dug out of the ground are destined for industry, to give cutting and drilling tools super-hard surfaces and sharp edges. The remaining 20% are some of the most perfect of nature's creations, all the colours of the rainbow and each unique. They are used to make some of the most expensive jewellery money can buy.

The hunt for diamonds is long and hard, but the rewards can be great. Treasures still reside deep in the Earth, it might just take a few million years for them to be brought to the surface and then found. It is predicted that some Australian deposits will run dry by 2007, but the new deposits in Canada look set to continue to produce diamonds for at least the next two decades. The search continues for the kimberlite pipes yet to be discovered.

CHAPTER THREE

The Cut

DIAMONDS IN THEIR ROUGH FORM ARE A FAR CRY FROM THE SHINING JEWELS

ON THE RED CARPET. SOME ARE BULKY AND MISSHAPEN, OTHERS SMALL AND JAGGED.

ABOVE The Hylle Jewel is a testament to the craftwork that flourished in the fourteenth century, particularly in the royal courts of France.

RIGHT The Imperial Orb of Emperor Matthias (1612–1619) from Austria, eye-catching with its diamonds, rubies, pearls and single sapphire.

The full extent of the beauty of diamonds only comes to light in the hands of expert cutters, who shape and polish the stones, turning them from rocks into riches. It is this process that brings out the 'fire' in a diamond: its amazing ability to reflect light within itself. Hundreds of millions of diamonds are mined every year, but only one in five is of good enough quality to be made into jewellery. And the better the cut given to these gem-quality stones, the more beautiful – and expensive – they become. It is worth the effort as the worldwide diamond jewellery market is worth around £28 billion (US$56 billion) per annum.

Cutting diamonds was once taboo. Early Indian manuscripts written more than 1500 years ago made it clear that although rough diamonds that sparkled and had a symmetrical shape could bring power and happiness, they must never be cut. The reluctance to shape or polish diamonds because of religious or other prohibitions eventually spread to Europe, and many of the early stones were kept as lucky charms.

It is only during the past 1000 years that diamonds have been more widely worn as jewellery in Europe, and only in the past few hundred that they have been worn by people other than nobility. By the thirteenth century, diamonds gradually began appearing in royal crowns, but these stones were small, low-key affairs and rarely the centrepiece, which was generally another stone such as sapphire.

No metal or other stone is hard enough to cut diamond. It can only be cut by another diamond, and most recently, by lasers. No one knows exactly how or where cutting started, but very early evidence of cut diamonds have been found, dating from about 1330, in Venice, which for years was one of the key trading points for the stone. As trade routes began to open up, more diamonds from India and via the Middle East started appearing in Europe, and by the sixteenth century the Belgian city of Antwerp had claimed the crown of diamond trading centre of the world. To this day Antwerp is one of the largest trading centres for

diamonds, with 80% by value of the world's rough diamonds passing through the city at some stage. Slowly diamonds made a name for themselves as desirable, beautiful stones any noble would be eager to own.

As the tools for cutting diamonds were developed, and as more diamonds were being found, so the gems as we know them today started to take shape. Up until the fifteenth century most diamonds were roughly shaped with a chisel and mallet, a primitive procedure that in less than expert hands would shatter the stone. The crude shaping process was limited to the natural shape of the diamond crystal, at best producing lumpy-looking diamonds with limited sparkle.

BELOW Diamonds have long been associated with love. This wedding ring c.1575 shows an Italian central motif of hands clasped in faith.

The process advanced in the fifteenth century with the invention of a manually operated polishing wheel, embedded with diamond dust. With the use of olive oil as a lubricant, the wheel could grind flat symmetrical shapes called facets on the cut diamond crystal. The popularity and demand for diamonds increased as cutting methods became more precise. By the eighteenth century, when Indian supplies had dwindled and new diamond resources had been unleashed in Brazil, and then in the nineteenth century

BELOW In a reference to Cupid, arrows pierce the heart at the centre of this elaborate brooch of diamonds, rubies and emeralds. It dates from c.1610–1620 and was probably made in Prague.

in South Africa, the advances fuelled the diamond business to trade more, sell more, cut more. As diamond trading increased across Europe, so the skills of cutting them grew. It wasn't long before diamonds were being sculpted for lavish settings in brooches, necklaces and rings.

Like any great painting or sculpture, a well-cut diamond is a work of art. Although the process has been improved by technical advancements, manual operations are still very important. Small stones are turned out in their thousands, but the bigger stones need the hands of a skilled master craftsman. Before a single blow is made to a rough diamond it must be examined in minute detail under a magnifying glass. An experienced cutter will want to consider the shape and position of 'flaws' before beginning to shape a stone. Flaws can be anything from large cracks to tiny crystals or dark spots. Some of these are known as 'inclusions', tiny specks

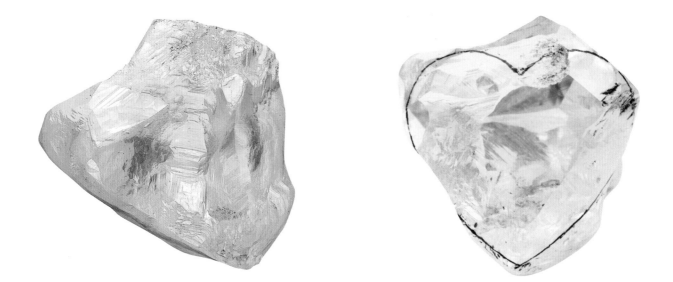

of minerals, sometimes including diamond, trapped inside the stone when it was formed. These are highly useful to scientists studying the origins of the diamond, but less useful when trying to create a perfect piece of jewellery. Most often, the fewer flaws a stone has the more valuable it is, so a cutter will painstakingly try and shape the stone by removing only the flawed parts.

The first step in cutting most diamonds is 'cleaving', where the stone is split into the best rough shape for its size. All diamonds have 'planes' along which the cutter makes tiny grooves with lasers or diamond saws. A sharp hit to these grooves will break the diamond along the plane, and so the cutter slowly works away at the diamond until the best part of the stone is left, ready to be shaped in more intricate detail.

ABOVE AND RIGHT A rough gem-quality diamond is first divided along its 'planes', before being shaped by a saw and 'ignited' by numerous reflective facets. The resulting cut diamond featured here is known as La Luna, created by Steinmetz.

LEFT Master craftsmen can take months to cut a stone. A wrong move could shatter it. The right one could double its value.

After cleaving, the stone is cut using precise diamond-edged saws or lasers, to further develop the shape before bruting, the process of rounding the stone to create its outer edge or girdle. As diamond is the hardest natural substance on Earth, only another diamond is durable enough to do this. The stone being shaped is cemented into a holder that fits on a turning shaft. Another diamond is cemented to the end of a long rod, which is used to grind the diamond into shape as it turns in the holder.

When dealing with one of the most expensive materials on Earth in this way, it is important that nothing goes to waste. The diamond chips worn away are collected and used for more sawing and shaping. Some small chips are sent to India, which specialises in cutting tiny fragments for uses like decoration around watch faces. If the chips are too small even for such

LEFT Such is the skill of a
cutter, diamonds as small
as strawberry seeds can
be crafted.

purposes or are flawed, they are used either in industry or ground down further into diamond dust, which is used to coat the rotating wheel used for the bruting and faceting stages.

Half a diamond's weight might be lost on the cutting floor, but this makes what is left all the more valuable. Once the best part of the diamond has been cut, then comes the magic. Using the diamond-coated wheel, the cutter grinds dozens of tiny flat surfaces or 'facets' into the stone's surface. Each of these facets acts like a little mirror, bouncing light around within the stone and bringing it to life. The cutter will use expert judgement to decide how large or small a particular facet should be cut to maximise its fire and sparkle. Decisions like this can make a difference of hundreds or possibly thousands of pounds.

Shaping a diamond in this way can take anything from under an hour to more than a year, depending on the size of diamond and quality of cut. Cutters must have good eyesight to make the delicate cuts and gentle shaping. Smaller stones are particularly difficult to shape, as the cutter often has to make minute cuts or grinds with the machine, then whip the diamond out of its holder to look at it under a microscope, before placing it in the holder again, to continue the slow, painstaking process.

The fashion for cut diamonds has changed over the years. Early stones from before the 1400s were 'point cut' to produce a basic octahedron shape with eight sides, like two pyramids stuck base to base. This was replaced by the 'table cut', which involved cutting the top of the point cut to create a flat surface. Table cuts were much brighter than points, a characteristic that soon began to shape the way for new diamond cuts. 'Rose cuts' with triangular facets became popular in the sixteenth century as the quality of a diamond or its colour increased in importance. Although bright, the rose still lacked a certain flair.

RIGHT This seventeenth-century painting of Princess Margherita Gonzaga from Italy shows her adorned with pre-brilliant cuts, from triangles and points to hearts and square tables.

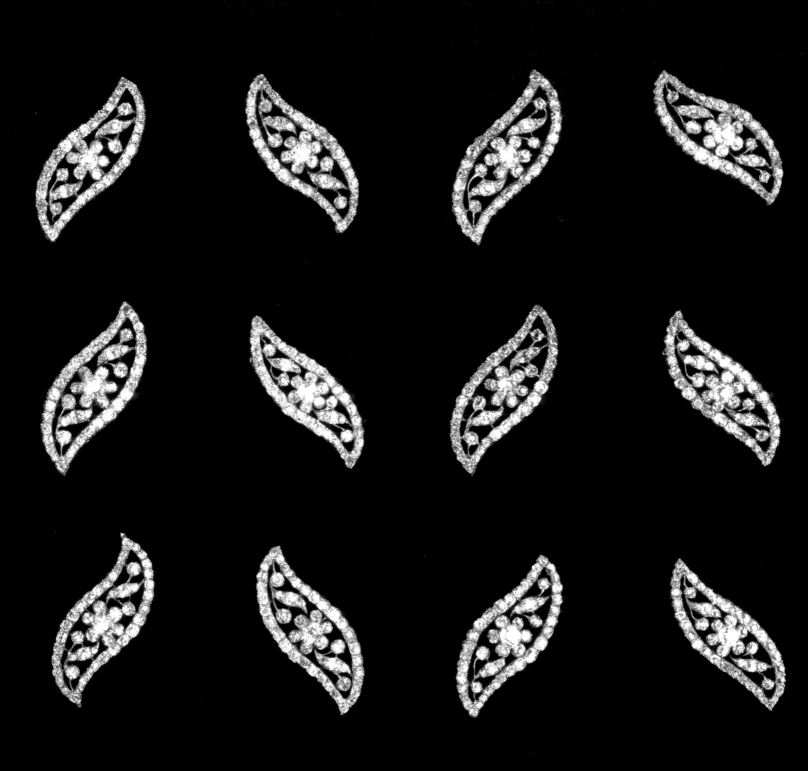

LEFT Formerly part of the Russian Imperial Collection, these brilliant-cut dress ornaments from around 1770 would have been stitched onto clothes.

Every diamond needs to make an 'entrance', and so as cuts evolved, more attention was given to how to maximise its shine. It wasn't until the 'brilliant' cut was devised in the seventeenth century that diamonds started to command the audience they deserved. By adding numerous facets to the flat-topped table cut, the stone started to reflect light like never before, and by 1920 the cut had been refined with mathematical precision so that the angle and number of facets maximised its 'fire'. That cut was called the 'round brilliant'. With 57 or 58 facets, this is the cut given to most modern diamonds and one that guarantees a diamond will get noticed.

Rough diamonds are cut all over the world today. Just as different deposits produce particular grades or colours of diamonds, certain countries have particular renown for diamond cutting. India has become a recent benchmark, and with its low-labour costs cuts nine out of ten of the world's diamonds. In terms of value, this represents 55% of the uncut diamond market, as India specialises in smaller, lower-quality stones.

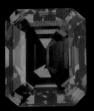

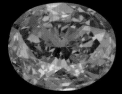

Israel is top of the high labour-cost league, exporting £3.2 billion (US$6.3 billion) worth of polished diamonds in 2004. It is here, as well as New York and Belgium that the larger, more difficult to cut and indeed more valuable stones are sent. Other cutting centres exist in China, Thailand and Sri Lanka. It is estimated that a total of about 900 million diamonds are polished every year worldwide.

In the past, the majority of the world's rough diamonds were sold through a single channel. However, growth in diamond production such as in Canada, as well as the interests of various governments in marketing their own production, has led to a multi-channel system. Today, the largest source of rough diamonds by value is the Diamond Trading Company, the sales and marketing arm of the De Beers Group. All companies selling rough diamonds set their prices using supply and demand statistics, the current state of the market and the likely future climate.

The Diamond Trading Company supplies diamonds to some of the world's leading diamantaires, known as 'sightholders'. Most sightholders

LEFT Variations of brilliant cut from top: round, radiant, marquise, oval and princess.

RIGHT Sorting or grading of diamonds happens at every stage from mine to shop. They are divided by size, shape, quality and colour.

are involved in diamond cutting and polishing in the world's major cutting centres, whilst others act as dealers. Rather than sell its diamonds by country of origin, the Diamond Trading Company takes in production from various sources, offering a wide-ranging 'mix' to its clients. It also sells industrial diamonds to a handful of clients. Sightholders initiate the distribution of diamonds down the supply chain - to the major diamond clubs (or bourses) that trade in uncut and polished diamonds. These are dotted around the world, mainly in Antwerp, London, New York, Tel Aviv, Mumbai and the Far East. The chain continues through polishing factories and jewellery manufacturers until the finished product reaches the consumer.

The final price of a gem-quality diamond is influenced by four key criteria, known as 'the four Cs': cut, clarity, carat, and colour. The quality of the cut rests with the cutting centre to which the stone is sent. The clarity

reflects the number of flaws or foreign particles, and each stone is weighed in carats, where one carat is precisely 0.2 g (0.007 oz). Colourless stones are the most popular. How colourless they are is graded in descriptive historic terms or by a lettering system from D to Z, where D is colourless and Z is a tinted colour. Intensely coloured stones known as 'fancies' come and go in popularity, but they are far more rare, sometimes caused by naturally included elements in the diamond's structure. For example, boron makes diamonds blue, and nitrogen is the cause of the yellowish tints in some stones. Red diamonds are the result of structural deformities and, along with blue, are the rarest of all. One of the causes of green diamonds is exposure to radiation in the Earth's crust. Black diamonds also exist, the most famous being the Black Orlov. Inclusions and defects can reduce a diamond's clarity so much they appear black.

The market price for coloured diamonds is heavily influenced by fashion. If black diamonds are in fashion this year, their price will go up. If people want square pink diamonds then their price will rise. The main buying influence is probably the United States, which consumes half the world's gem-quality diamonds. Japan is the second biggest consumer with large markets elsewhere in the Far East and Middle East. Marketing schemes to relaunch brown diamonds as 'champagne' or 'cognac' diamonds have worked to an extent, with Australia being the biggest producer of them globally. Australia is also renowned for its small but consistent production of exquisite and rare coloured diamonds – particularly, pink, red and purple ones – which have found sizeable markets.

These 'cognac' diamonds are from the famous Aurora Collection, the largest collection of coloured diamonds in the world, comprising 296 stones.

The cutting of a diamond is a crucial landmark in its history. Once cut, a stone's origin cannot be easily traced, making it easier to sell on as a conflict diamond. Some diamond companies in Canada engrave their gems with a microscopic tag such as a polar bear. These are invisible to the naked eye, but guarantee customers the stone is of Canadian origin and cut and polished to the world's highest standards.

In addition to these kinds of initiatives, retailers and consumers should ask about the origin of the diamonds and company policies in place to prevent dealing in conflict diamonds. The Diamond Trading Company has introduced the 'Forevermark', a mark of trust, which is a microscopic inscription placed on the table of the stone. It guarantees the diamond is genuine, natural and supplied by the Diamond Trading Company. The Forevermark also adds reassurance to the consumer that the stone has not been artificially improved by using heat and radiation to clear impurities and add colour. It is clear that diamond buyers have more than just the stone's cut to consider.

TOP AND ABOVE
Canadian diamonds from the company Sirius are branded with a polar bear.
The Diamond Trading Company inscribes its diamonds with the 'Forevermark'.

CHAPTER FOUR

Fame

DIAMONDS AND FAME GO HAND IN HAND. SOME STONES ARE FAMOUS

FOR THEIR SIZE OR COLOUR, OTHERS FOR THE PEOPLE THAT WEAR THEM.

Many diamonds are renowned for their high value, and a few for the stories of their past. But often that fame comes at a price, where they attract the eye of the thief as well as the admirer. Some of the most legendary stones are from India and the most famous of these is arguably the ill-fated Koh-i-Noor. With a name meaning 'mountain of light' in English, the stone in its old 'Indian cut' was proudly worn by royalty for hundreds of years. In 1849, the British demanded it as a tribute, following the ruler of the Punjab's defeat in the Sikh wars. It was then presented as a gift to Queen Victoria a year later. It now sits re-cut as a dazzling 109-carat stone embedded in the Queen Mother's crown in the Tower of London. The value of the Koh-i-Noor is impossible to guess, although a similar-sized stone sold for £8.3 million (US$16.5 million) in 1995, a world record for the sale of a single diamond.

Diamonds are famously expensive, and the most costly piece of jewellery ever sold was a pair of diamond drop earrings, bought anonymously for £2.9 million (US$5.8 million) at Sotheby's Geneva in 1980. Jewellery of this value is most likely to be kept under lock and key, but occasionally they appear in the limelight. Elizabeth Taylor's love affair with diamonds is no better expressed than by a stone she asked her then husband Richard Burton to buy in 1969. The show-stopping 69-carat diamond went to auction, but Burton narrowly lost out to Cartier in the sale, who secured it for just over £500,000 (US$1 million). Knowing Taylor's desire for it, Burton called Cartier from a hotel pay phone later that day and negotiated a private sale of the stone he later named the Taylor-Burton. The conversation must have made for irresistible eavesdropping with residents of the hotel.

LEFT The Taylor-Burton diamond, formerly known as the Krupp.

RIGHT Elizabeth Taylor wore the famous stone to the 42nd Academy Awards in 1970. She sold it eight years later following her divorce from Richard Burton.

But not all diamond stories have a happy ending. The stone that stirs the most intrigue and caution is rather inappropriately called the Hope, named after one of its owners, Henry Philip Hope. The beauty of this blue 45.5-carat stone belies the apparent curse on it, which promises anything from bankruptcy to death to its luckless owners. These legends were no deterrent to a strong-headed and very wealthy American, Mrs Evalyn Walsh McLean, who in 1910 bought it as a 'lucky charm'. Whether the legends were true or not will never be known, but it is true that it was McLean's tragic fate to see her nine-year-old son die in a car crash, suffer the suicide of her daughter and watch helplessly as her husband spent the last of his days in a mental institution. McLean was the last of the Hope's private owners. In 1949, two years after her death, jeweller Harry Winston bought it before presenting it to Washington's National Museum of Natural History ten years later, where it continues to be one of their most popular exhibits. Researchers have only recently been able to use computer analysis to trace its origins. The Hope had apparently been cut from a larger stone, known as the French Blue, which was once part of the crown jewels of France, and had disappeared during the Revolution of 1792.

LEFT Evalyn Walsh McLean wearing the Hope as a pendant. This extravagant diamond appealed greatly to her flamboyant character.

RIGHT The Hope is perhaps the best recognised gem in the world.

The biggest rough diamond ever found was the magnificent Cullinan. Unearthed in South Africa in 1905, it weighed a staggering 3106 carats, or 0.6 kg (1⅓ lb) – about the weight of four snooker balls. Artfully cut into nine large gems and 96 smaller stones, it was the source of the 530-carat Great Star of Africa, the largest colourless cut diamond in the world and now in the Sovereign's Sceptre as part of the British Crown Jewels. It joined the ranks of the world's most beautiful big diamonds, including the Black Orlov and the Regent from India. In more recent times, the fabulous Centenary was unveiled by De Beers on the company's hundredth birthday in 1988. This clear, colourless gem is a stunning piece of work, with 164 facets around the top of the stone and 83 on the girdle.

LEFT Also known as the Cullinan I, the Great Star of Africa now sits in the Sovereign's Sceptre with Cross.

RIGHT The rough Cullinan, held here by a miner in South Africa, was the size of a fist.

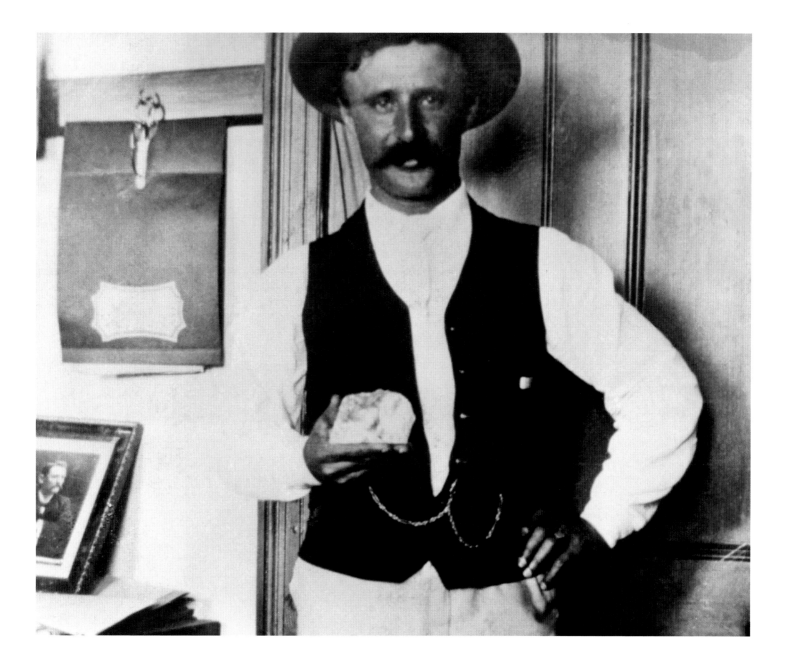

RIGHT The colouring of the
Dresden Green is caused by
natural radiation.

The even rarer coloured or 'fancy' stones are among the most prized and famous. It took 20 years for the sensational Aurora Collection to be put together, now comprising 296 diamonds of a whole range of different colours. The jade-coloured Dresden Green, at 41 carats, is one of the few naturally green, large diamonds in the world. And of course there is the iconic Tiffany, a stunning yellow, 128-carat, 90-faceted stone owned by the famous New York jewellery house since 1879. Today, it features in a stunning setting of a 'Bird on a Rock' designed for Tiffany in the early 1960s.

BELOW The famous 'Bird on a Rock' brooch is on permanent display at Tiffany in New York.

One of the most recent large diamonds to be found is famous for reasons beyond its beauty and immeasurable worth. The De Beers Millennium Star was discovered in 1992, in what is now the Democratic Republic of Congo. An astoundingly beautiful 203-carat flawless pear shape, it was an irresistible target for criminals when in 2000 it was displayed in London's new visitor attraction, the Millennium Dome. One morning, thieves drove a digger at high speed through the site's perimeter fence towards the De Beers Millennium Star and the eleven other diamonds on display. Thanks to a tip-off, they were intercepted by undercover police posing as cleaners and other staff. Even if the robbers had made it to their getaway speedboat, they would have found their stash consisted merely of crystal replicas planted the day before by police and worth only a fraction of the £350 million (US$700 million) the real gems could fetch.

Not all diamond thefts are diverted. In Paris in 2003, thieves helped themselves to two diamonds together worth more than £7.8 million (US$15.6 million). And the year before, an Italian gang raided some of the vaults under the Belgian city of Antwerp and emptied 123 of 160 safety deposit boxes. The estimated booty was worth £65 million (US$130 million), but as many of the box owners have been reluctant to come forward and identify what was stolen, it may never be known how much was lost.

RIGHT Master craftsmen removed 75% of the rough diamond to produce the flawless De Beers Millennium Star, created by Steinmetz.

For most, the fame of diamonds rests with those who wear them. At film premieres or high-profile charity events it is often the jewels and dresses that make the headlines. Halle Berry wore £2.5 million (US$5 million) worth of blue and white diamonds to the 60th Golden Globe Hollywood film awards in 2002, a modest outfit when compared with the shoes made by top shoe designer Stuart Weitzman, comprising 464 diamonds. And who could compete with the most expensive swimsuit in the world? Making a splash at New York Fashion Week 2005, it was adorned with £9 million (US$18 million) worth of diamonds surrounding a rare 103-carat Golconda diamond.

Top-class jewellers such as Van Cleef & Arpels, Tiffany and Cartier vie with each other for this kind of publicity. Boodle & Dunthorne benefited hugely from the coverage and subsequent business it received having supplied David Beckham with his engagement ring and trademark diamond stud earrings. Celebrities 'wearing' their wallets like this, be it diamond bikinis or blinding jewels, spawned the 'bling' culture, a product of the past few years that is losing its lustre. Bling is about having big cars, big jewels and even bigger bank accounts. It has been seen as an expression

LEFT The most expensive swimsuit in the world was revealed at New York Fashion Week in 2005, priced at £9 million (US$18 million).

RIGHT The ultimate bling motif, a dollar-sign pendant encrusted in diamonds.

of the best money can buy. Many of the bling ambassadors are in the music business. Hip-hop artist 50 Cent gave his fans a chance to enjoy a bit of bling for themselves when he released a CD called *Beg for Mercy*. Four copies of the CD enclosed a ticket entitling whoever found it to a £7000 (US$14,000) diamond medallion.

But of all the popular uses of diamonds, the engagement ring is the enduring success. The idea that everyone could own a diamond, symbolising love and eternity, was the brainchild of De Beers, which launched its 'a diamond is forever' campaign in 1947. Today, hundreds of thousands of diamond engagement rings are sold every year across the world. And when celebrities get engaged, the story isn't complete without a full debrief on the details of the ring. Singer Jennifer Lopez and actor Ben Affleck's engagement sparked a rush to buy pink diamonds after he presented her with a £500,000 (US$1 million) pink diamond ring. Although not all engagements have the longevity characteristic of a diamond, it would seem this tradition is the hardest to break – does a ruby or sapphire say 'I love you' as much?

LEFT This pink diamond is from the Aurora Collection, one of the finest and most comprehensive collections of naturally coloured diamonds in the world.

CHAPTER FIVE

Extreme

DIAMONDS MAY ADORN THE RICH AND FAMOUS, BUT LOOK BEYOND THE RED CARPET

AND THERE'S MORE TO THEM THAN SPARKLE. DIAMOND IS EXTREME IN EVERY RESPECT.

D iamond is the hardest natural substance on Earth, harder even than steel. Unlike other minerals, it is unaffected by chemical attack and won't disintegrate after long periods of time in water or acid. Diamond can survive intense heat, high pressures and radiation. For years it has been extracted from the sands and gravel beds of rivers, and now scientists have found tiny crystals of it in space.

Diamonds that are too flawed or too unfashionably coloured are used in the manufacturing industries. In fact, 80% by volume of the world's diamonds are of industrial quality, exploited for their supreme properties.

Because diamond is so durably hard, it is now the material of choice for pinpoint cutting, grinding and polishing tools. Stonemasons' saws are lined with tiny crystals that make it easier to carve out blocks of marble and sandstone. Surgeons use knives tipped with a single diamond crystal to slice through soft tissue with phenomenal accuracy. Even dentists now use disposable diamond drill tips. Diamond tools are also used for high-level polishing, from pistons in a car engine to delicate contact lenses.

RIGHT Rough diamonds hold huge potential. Most are used by industry for superlative cutting tools and as heat spreaders.

The electronics industry values diamond for its ability to transfer heat more efficiently and faster than any other material. Diamond crystals are added to copper wires to increase the rate at which heat is transferred through them, preventing overheating and circuit breakdowns.

But using industrial diamonds on a large scale is costly. For years, manufacturers could only dream of a substance with the magnificent properties of diamond but a more humble price tag. The dream came true in the 1950s, when independently the American company General Electric and the Swedish company ASEA did the impossible and found a way to make diamonds from scratch. They were able to grow single diamond crystals from graphite, another naturally occurring form of carbon much softer than diamond. With intense heat and pressure, they created in minutes the Earth's natural process of making diamonds. Industry now uses 20 times as many synthetic as natural diamonds.

This method of growing diamonds revolutionised the manufacturing business. More and more industries could afford to harness the superlative properties of diamond in their products. Crushed industrial diamonds were added to rubber, reducing the likelihood of splitting and wear. Diamond paints gave a formidable resistance to acids, damp and abrasion, and diamond was also added to some skin exfoliants. Even the average golfer benefited, where new super-hard diamond hitting surfaces on clubs transfer all the force of the hit to the ball, with virtually none absorbed by the club.

Diamond's extreme transparency to light makes it invaluable to industry and science. It lets a wide spectrum of light, from ultra-violet to infrared, pass through, unlike much glass, which is transparent to visible light only. The first investigative space probe sent to Venus in the 1970s made ingenious use of this particular quality of diamond. Individual crystals were used to create tiny transparent 'windows' between sensitive detectors onboard the probe and the harsh environment outside. These windows allowed signals from the surrounding environment to pass easily through to the detector. They were also unaffected by the extreme conditions outside, and so were resistant to becoming scratched or obscured. Diamond windows are also used in missile heads that need to pick up infrared signals.

As more and more industries used diamonds, there was a drive to find new ways of making them. The high-pressure method of growing individual crystals was laborious, and produced only relatively small diamonds. Research into a dramatically improved process began to gather momentum 20 years ago, which was to spell the future for synthetic diamond material. Known as chemical-vapour deposition (CVD), it produced clouds of microscopic diamond crystals from carbon-containing gases such as methane. This mist of crystals can be used to drape solid objects like razorblades or drills in a hardwearing diamond coat. CVD diamonds are also not restricted to crystals, and can be grown in sheets or moulded shapes.

This new way of making diamonds has meant its use is now almost limitless. Experts predict CVD diamond will revolutionise the future of computers in its ability to transfer heat. As designs improve and modern machines get smaller, heat transferral becomes a major concern. Silicon is currently the main component of electronic systems, and more and more parts are today coated in CVD diamond. But imagine using tiny squares of CVD diamond film instead. Able to withstand five times as much heat as silicon and far less likely to wear out, diamond may be the future for high-frequency, high-voltage applications, such as satellites and power grids. CVD diamond could also bring us cheaper and more efficient flat

LEFT Future nanorobots with diamond parts will be small enough to work in the bloodstream, drilling clots or injecting infected cells, as this illustration shows.

screens, where diamond is used to generate the charged particles needed to produce the image, but at a much lower voltage. Diamond windows that were once restricted to the size of individual crystals can now be made of a thin sheet of CVD diamond and so can be larger.

Diamond also has potential application in surgical treatments, as due to its extreme durability and chemical resistance it can be used in the human body. In the future, CVD could be used to coat joint replacements in patients, to save them being replaced after 15 years or so. It is also predicted that before long we will have machines with diamond parts so small they can be injected into the bloodstream and programmed to monitor blood sugar levels in diabetics or clear a blocked artery. The problem with making moving parts for these 'biosensors' out of almost any material is that the friction they experience, as they rub against one another, would soon wear away such minute pieces entirely. Intense research is underway to find ways of making them from uniquely tough diamond.

The ingenious and at present 'extreme' uses of diamond touched on previously will affect all our lives in the future, from our technology to our health. Scientists are even working to grow new nerves on diamonds. In 2002, one consortium of researchers in the United States embarked on an ambitious plan for visually impaired people suffering from diseases such as retinitis pigmentosa and macular degeneration. They patched the damaged retinas with diamond-coated microchips. These take the place of the damaged rods and cones that normally convert light into electrical signals to send back to the brain, helping the person regain a degree of sight.

The technological future for synthetics is bright. But what are the implications for the diamond trade? When synthetic crystals were first produced 50 years ago, they were very small and obviously flawed with the telltale signs of manufacture. Today, it is possible to grow one-carat synthetics of gem-quality. Of the two companies that sell gem-quality synthetics today, one offers 'fancy' yellow and orange stones, making virtue of imperfections in the process that causes nitrogen to colour them. The other company produces what it calls 'nearly flawless' synthetics – ones with tiny imperfections. The overall volume of synthetic stones produced a year is very small and all are easily detectable. However, in the future it may be possible for perfect, flawless synthetics to be made to order.

The real test for synthetics will be in the marketplace. The price of natural diamonds is hugely dependent on the desire to own the ultimate in luxury, a stone formed in the depths of the Earth billions of years ago. Marketing made-to-measure diamonds born in a laboratory might be more difficult. Such a token would certainly seem a lot less romantic in sentiment. It would be the same as trying to market a mimic of a luxury sports car. It may look like a Porsche and drive like a Porsche, but if there

ABOVE AND LEFT
Gem-quality synthetics can now
be made to exquisitely high
standards, such as those featured
in the bracelet shown here.

were thousands of others like it going cheap, would it still feel like one? One offshoot market for synthetics is offered by an American company called LifeGem, which will take the ashes of your dead loved ones and then compress the carbon elements to form a diamond. At up to £11,000 (US$22,000) per stone, it brings a new meaning to the term 'family jewels'.

The existence of diamonds extends beyond the boundaries of our atmosphere into outer space. They have been found in a small number of the meteorites which fall to Earth each year. Inside some of these diamonds scientists have discovered yet more treasures. Rare gases trapped within reveal they formed in stars which existed before the Solar System. These diamonds, along with other material coalesced to create the Solar System and its constituent planets, billions of years ago.

The drama of this superlative gemstone is no better expressed than by a remarkable discovery in 2004 in outer space. Astronomers found an enormous chunk of diamond, an astonishing 4000 km (2485 miles) across. The compressed heart of an extinct star, it once shone as brightly as our Sun. Extreme indeed.

CHAPTER SIX

Design

THE MOST ALLURING AND FAMILIAR EXPRESSION OF A DIAMOND'S VERSATILITY

AND BEAUTY LIES IN THE BREATHTAKING JEWELLERY IT IGNITES.

Single stones may have a romance so typical of diamonds, but for many it is the diamonds set into jewellery that have the most dramatic effect. Whether these are ribbons of diamonds draped into a necklace, or explosions of coloured stones in a tiara, such pieces are unmatched in appeal. Famous diamond wearers such as Wallis Simpson, wife of the abdicated King Edward VIII, found it hard to resist the charms of diamond jewellery and began collecting it. Her glittering stockpile sold for £31 million (US$62 million) in 1987 and included some spectacular pieces. The classics of diamond jewellery, with some careful handling, will be with us for years to come, and antique pieces are as fashionable now as they ever were. Diamond settings from the eighteenth century were particularly dazzling bows and bouquets, tiaras, brooches, buttons and earrings. With such scope, what are the inspirations for contemporary jewellery designers?

LEFT This brooch featuring the Williamson diamond was created by Cartier at the request of Queen Elizabeth in 1953, the year of her coronation.

RIGHT A breathtaking wreath tiara of brilliant-cut diamond flowers and ruby stamens c. 1835.

LEFT The pear-shaped Tiger Tears have world-ranking status. Each weighs more than 50 carats.

RIGHT A striking white, heart-shaped diamond ring with diamond-studded band created by Graff. The total diamond weight is 4.28 carats.

The future for diamond jewellery is already here. Sponsored diamond competitions help encourage upcoming designers to keep the stone's appeal at the cutting edge of the market. The Diamond High Council Awards in Belgium began in 1984, a bi-annual diamond jewellery competition run by the Antwerp Diamond High Council to stimulate creativity and promote contemporary designs. With almost 500 entries in 2004, this is obviously an important outlet for tomorrow's designers. Open to professional and non-professional jewellery designers, students, goldsmiths, silversmiths and designers in general, the competition aims to enhance creativity and excellence in contemporary diamond jewellery. Winners are chosen for the creativity of their design, its aesthetic quality, wearability and craftsmanship.

The Diamond Trading Company also run a similar competition for young South Africans, helping to promote the country's natural heritage with innovative design. South Africa's reputation as the producer of some of the world's finest diamonds can only benefit by encouraging its national designers to incorporate these gems into artworks.

FAR LEFT 'Taraxacum' ring by Marleen Schodts, 2003 finalist in the Antwerp Diamond High Council Awards with the theme 'movements'.

LEFT 'Dancer' bracelet by Alessandro De Paula Alvarenga, finalist in the Antwerp Diamond High Council Awards 2003 with the theme 'movements'.

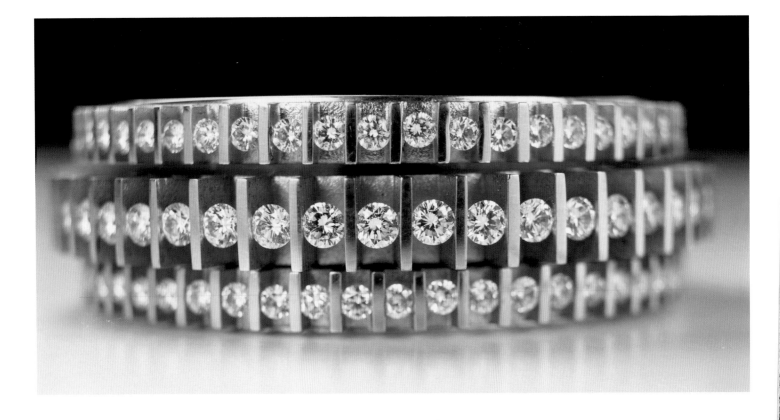

FAR LEFT 'Turbine' bracelet by Cari-Mari Wilsenach, 2003 finalist in the Antwerp Diamond High Council Awards with the theme 'movements'.

LEFT 'Hidra' dog collar by Fernando Maculan and Adriano Mol, 2003 winner of the Antwerp Diamond High Council Awards with the theme 'movements'.

TOP Innovations for diamond design have spread beyond rings and earrings, to teeth.

ABOVE A diamond eye adorns the logo on the Jaguar F1 racing car at the Monaco Grand Prix.

For much of the twentieth century, diamonds were predominantly a woman's domain. Jewellery has now become just as much a male expression of individuality and status, which has helped steer diamond design to ever more elaborate ends. Even the most conservative man can wear diamonds without feeling self-conscious. Cuff-links, lapel pins, neck-tie clips, belt buckles and wallets can all be enhanced with a discreet diamond here or there. And for the more flamboyant, no stone is better suited to showing off. Chunky rings, diamond pendants and thick bracelets are all on offer.

Diamond designs are not limited to jewellery. In 2004, Jaguar embedded £140,000 (US$280,000) worth of diamonds into the nosecones of two of their Formula One racing cars, to promote the film *Oceans 12* starring George Clooney and Brad Pitt. When this was revealed at Monaco's Grand Prix, Jaguar snapped up the headlines, but only because one of the diamonds was lost.

The lustrous appeal of diamonds cannot be disputed. The as yet unrealised market of synthetics may find a place in innovative design, so who knows what tomorrow will bring. Diamond chairs and tables, diamond knives and forks, or even diamond corkscrews and mobile phones? Maybe. With a mineral so versatile, anything's possible.

Index

Picture credits

Cover image, end paper, chapter openers, p.7 © De Beers LV; p.8 © V&A Images/Victoria and Albert Museum; p.10 Photo RMN © Thierry Ollivier; p.11 Cartier Archives © Cartier; p.13 Tiffany&Co. © Stephen Lewis; pp.14-15 © GRAFF DIAMONDS; pp.16-17 © Diamond Trading Company Limited; pp.18, 22, 27 © De Beers; p.28 © COLLART HERVE/CORBIS SYGMA; p.31 © Peter Johnson/CORBIS; p.32 © De Beers; p.35 © Diamond Trading Company Limited; p.36 © Gallo Images/CORBIS; pp.38-39 © Diamond Trading Company Limited; p.42 © New College, Oxford/ Bridgeman Art Library; p.43 © Kunsthistorisches Museum, Wien; pp.44-46 © V&A Images/Victoria and Albert Museum; pp.48-49 © Steinmetz Diamond Group; p.50 © Charles O'Rear/CORBIS; p.52 © Cary Wolinsky/National Geographic Image Collection; p.55 © Courtesy of the Ministero dei Beni e le Attività Culturali; p.56 © V&A Images/Victoria and Albert Museum; p.58 top © Aurora Gems/Photo Tino Hammid, 2nd from top © Aurora Gems/Photo Robert Weldon, middle © Diamond Trading Company Limited, 2nd from bottom © Aurora Gems/Photo Robert Weldon, bottom © Diamond Trading Company Limited; pp.59-61 © Diamond Trading Company Limited; p.62 © Chuck Mamula Photography; p.63 top left © Aurora Gems/Photo Robert Weldon, top right © Aurora Gems/Photo Tino Hammid, bottom left © Aurora Gems/ Photo Robert Weldon, bottom middle © Aurora Gems/Photo Tino Hammid, bottom right © Aurora Gems/Photo Robert Weldon; p.65 top © Reuters/CORBIS, bottom © Diamond Trading Company Limited; p.68 The Royal Collection © 2005, Her Majesty Queen Elizabeth II; p.70 Cartier Archives © Cartier; p.71 © Bettmann/CORBIS; p.72 Cartier Archives © Cartier; p.73 © Smithsonian Institution/Bridgeman Art Library; p.74 The Royal Collection © 2005, Her Majesty Queen Elizabeth II; p.75 © Hulton-Deutsch Collection/CORBIS; p.76 Staatliche Kunstsammlungen, Dresden; p.77 Tiffany&Co. © Craig Cutler; p.79 © De Beers LV; p.80 © Getty Images; p.82 © Petre Buzoianu/CORBIS; p.83 © Getty Images; p.89 © Diamond Trading Company Limited; p.92 © William Whitehurst/CORBIS; p.94 © Erik Viktor/Science Photo Library; p.97 © The Gemesis Corporation; p.98 © Canada-France-Hawaii Telescope/Jean-Charles Cuillandre/Science Photo Library; p.102 Cartier Archives © Cartier; p.103 © V&A Images/Victoria and Albert Museum; p.104 Cartier Archives © Cartier; p.105 GRAFF DIAMONDS; p.106 HRD Awards 2003, Marleen Schodts, Warcel Diamonds bvba © Diane Hendrikx; p.107 HRD Awards 2003, Alessandro De Paula Alvarenga, Windiam bvba, Forum Romano © Diane Hendrikx; p.108 HRD Awards 2003, Cari-Mari Wilsenach, Intralcor nv, Metal Concentrators © Diane Hendrikx; p.109 HRD Awards 2003, Fernando Maculan & Adriano Mol, Intergems-Claes nv, Talento Joias, Alexandre © Diane Hendrikx; p.110 © Michelle Elmore; p.111 © Getty Images.

All other images are © 2005 Natural History Museum, London.
Every effort has been made to contact and accurately credit all copyright holders. If we have been unsuccessful, we apologise and welcome corrections for future editions or reprints.